The Ambassador of Bourbon

Maker's Mark and the Rebirth of America's Native Spirit

Photographs and Text by
David Toczko

Foreword by
Rob Samuels

Introduction By
Bill Samuels, Jr.

Acclaim Press
MORLEY, MISSOURI

Acclaim Press
— *Your Next Great Book* —
P.O. Box 238
Morley, MO 63767
(573) 472-9800
www.acclaimpress.com

Book Design: David Toczko and Shelley Davidson
Cover Design: David Toczko, Doe-Anderson and M. Frene Melton

Copyright © 2012 David Toczko
All Rights Reserved.

Maker's Mark®, SIV®, the SIV Device, and the Dripping Wax Device are all registered trademarks of Maker's Mark Distillery, Inc. and are used with permission. Maker's Mark Distillery, Inc. does not sponsor or endorse *The Ambassador of Bourbon: Maker's Mark and the Rebirth of America's Native Spirit*.

No part of this book shall be reproduced or transmitted in any form or by any means, electronic or mechanical, including photocopying, recording or by any information or retrieval system, except in the case of brief quotations embodied in articles and reviews, without the prior written consent of the publisher. The scanning, uploading, and distribution of this book via the Internet or via any other means without permission of the publisher is illegal and punishable by law.

Library of Congress Control Number: 2012908762

ISBN-13: 978-1-938905-00-1
ISBN-10: 1-938905-00-8

First Printing: 2012
Printed in the United States of America
10 9 8 7 6 5 4

This publication was produced using available information.
The publisher regrets it cannot assume responsibility for errors or omissions.

Contents

The History of Maker's (As Told by Bill) 10

Arriving at Maker's 30

Margie's Kitchen 42

The Quart House 46

The Barrel Wagon 56

Stillhouse Exteriors 60

Cooking and Distilling 70

Quality Control 102

Making the Barrels 106

The People and the Place 118

Filling the Barrels 124

Off to the Rickhouse 130

Maker's Mark® bourbon is "Born" 146

Making the Label 150

Filling the Bottles 154

Let's Take a Dip 158

The Iconic Bottle 168

Let's Take a Taste 172

Whisky Creek 180

The Grounds 182

Seldom Seen Sights 192

Dedication

To Ashley and Christopher,
my children and the "Spirits" that fill my heart.

Acknowledgements

With a project of this scope, completing it could not have been possible without the advice, encouragement and support of Beth Roberts upon whom I rely and love so much. No truer case of "behind every man is a great woman" can be found than within these pages.

I would also like to thank the entire Maker's Mark family. They welcomed me into the fold and provided me both the opportunity and access to capture the images that help tell their story. A special thanks to Chairman Emeritus Bill Samuels, Jr. and Chief Operating Officer Rob Samuels. Their leadership, along with their involvement, support and vision helped make this book possible. Thanks also to Victoria MacRae-Samuels, Maker's Mark's Vice President of Operations, and Sydina Bradshaw, Director of Visitor Relations, for the time they took out of their busy schedules to help facilitate the project.

David Vawter and a host of others at Doe-Anderson not only provided input with regard to the book's layout, but also provided the archived photos contained herein.

Jimmy Wickham of Independent Stave Company has my gratitude for showing and explaining to me their barrel manufacturing process from which the captions came for that part of the book.

Foreword

I'm very pleased that you've taken the time to pick up this book on our whisky and the very special place where we make it. As the eighth generation of the Samuels family who've made their livelihood in the craft of making whisky, I can attest to the fact that what we do is a labor of love. And nowhere is that love more apparent than on the grounds of our little distillery.

When my grandfather, Bill Samuels, Sr., purchased the abandoned Victorian distillery that was to be the home of Maker's Mark, he didn't do it because of the beauty of the surroundings. He did it because it contained a plentiful source of iron-free water that he knew would make his whisky taste better. Also, in 1953 the place was, frankly, far from beautiful. It had been vacant for decades and showed it.

It was my grandmother, Margie Mattingly Samuels, who saw the potential in the weatherbeaten buildings and unkempt lawn to make the Maker's Mark Distillery a place as unique as the whisky that was being made there. So once the company began running – just barely – in the black, Margie asked Bill, Sr. (insisted is more like it) that a portion of the profits be reinvested in turning the distillery into the kind of place that fans of our whisky would want to visit.

Once she was satisifed with her efforts, we threw open the doors to anyone who was interested in seeing how we made our bourbon. In fact, my Aunt Leslie was the very first tour guide. More than anyone else's, the distillery reflects her love, unstinting energy and passion to create something that future generations of Samuels (and Maker's Mark lovers) could be proud of.

Thanks to her vision and hard work, in 1974 the Distillery was listed on the National Register of Historic Places and in 1980, it was declared a National Historic Landmark. Today it remains the oldest operating bourbon distillery in the world – and the one and only place where Maker's Mark is made.

I think you'll agree that David Toczko has done a beautiful job capturing the charm of the place, and it's my hope that it will inspire you to come pay us a visit. We'd love to show you around.

Sincerely,

Rob Samuels
Chief Operating Officer, Maker's Mark Distillery, Inc.

Preface

Nestled in the rolling hills of central Kentucky, just outside the sleepy little town of Loretto, lies the Maker's Mark® Distillery, Inc. Already declared a National Historic Landmark, it is also considered by many bourbon lovers as the Home Place of Fine Bourbon. I grew up in Kentucky and learned the story of Maker's Mark® bourbon along the way. I have also, on more than one occasion, enjoyed savoring the fruits of their labor. It was not, however, until 2009 that I made my pilgrimage to Loretto and joined the thousands of people who visit this special place each year.

I wandered the beautiful grounds. I took in the aroma of the mash wafting in the air. I saw the process of how the bourbon is made and met the people who make it. This was my "ah ha" moment and it was then that I decided to produce this book. It is presented as a pictorial walking tour and follows much the same path as any visitor would take (with a few special stops along the way). I hope that those who have not been there will be inspired to visit and those who have will enjoy the experience again.

In these days of mass production, automation and high tech, it's refreshing to see an example of doing it right, doing it the old fashioned way, doing it with pride and doing it by hand. One can't help but be struck by the blatant inefficiency inherent in such an operation, while at the same time be impressed by the value placed on tradition. I have attempted to portray that dedication, the emphasis on a handmade product and what makes Maker's Mark, Maker's Mark® bourbon, throughout the book.

I learned quickly that everyone at Maker's Mark takes their bourbon seriously, but not themselves. This is driven from the top down as Bill Samuels, Jr. obviously is not shy about poking fun at himself (a few examples are included). The Samuels family has a distilling tradition that dates back over eight generations. To fully chronicle the family's long and colorful past would require a work along the lines of War and Peace. For that reason, our look into the world of Maker's Mark will begin with Bill Sr.'s re-entry into the distilling fraternity in 1953.

Pre-History
175 Years of Making Whisky the Old-Fashioned Way
(Like Everyone Else)

"For the better part of two centuries, since Robert Samuels first put down roots in Kentucky, our family were whisky makers. Unfortunately, there was nothing particularly special about the whisky we made."

— Bill Samuels, Jr.
Former President and Chairman Emeritus

Upper left: *Back in the day, distillers were friends as well as competitors, and here's the proof. On the far left in the top row in this shot is Les Samuels. To his immediate left (to the right in the photo) is Jim Beam. Yes, that Jim Beam.*

Above left: *A collection of original bottles and decanters of various whiskies made by generations of Samuels.*

Above right: *Letterhead of the T.W. Samuels Distillery, 1897.*

Large photo: *Staff of the T.W. Samuels Distillery, sometime shortly after the end of Prohibition.*

Inset at right: *What whisky labels used to look like (before Margie Samuels got in on the act).*

Bill Samuels, Sr. Breaks with Family Tradition
(Actually, He Sets it on Fire)

"After coming back from World War II, my Dad spent a few years in the family business, and came to the conclusion that he didn't want to make any more average whisky. So with all of us kids looking on, he burned the old family recipe and set about creating his own. These are the kinds of things that make a real impression on you when you're a thirteen-year-old kid."

— Bill Samuels, Jr.
Former President and Chairman Emeritus

Left: *Maker's Mark founder T. William Samuels in uniform, sometime during World War II.*

Below: *The secret of our success: The unique mash bill, substituting soft winter wheat for the usual rye, gives Maker's Mark® bourbon its full-flavored yet easy-to-drink character.*

Bill Samuels, Jr., son of the founder, reenacting his father's burning of the old family whisky recipe in front of the Maker's Mark Stillhouse.

Top left, top right and above: *Rare early shots of the Maker's Mark distillery and its first employees.*

(handmade)
handcrafted
hand selected
homemade
family history

(Original)
excellence
blue ribbon
best
seal of approval
sign of excellence
(hallmark)
(signature)
one of a kind!
champion
gold
silver
pewter

hallmark
handmade
original
maker
signature
Hallmark Bourbon
Maker's Bourbon
(Maker's Mark)

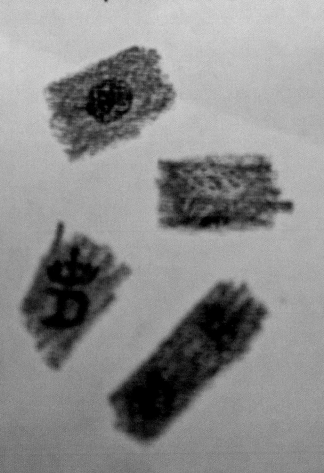

Mom Knows Best

"My mother, Margie, was not a trained marketer. Nor was she a graphic designer by trade. Despite these obvious handicaps, she named Dad's whisky, gave it its distinctive hand-dipped red wax top, created the typeface and the label itself.

Maybe there's a lesson in that somewhere."

— Bill Samuels, Jr.
Former President and Chairman Emeritus

Margie Mattingly Samuels, wife of Bill Sr., mother of Bill Jr.

Above: *Mom and Dad shortly after marrying, photographed in front of the old T.W. Samuels distillery.*

Right: *Mom front and center.*

A couple of Mom's big ideas: Hand-dipping every bottle in red wax; and the hand-torn label with Mom's typography, shown here on the very first bottle of Maker's Mark®.

From Genteel Obscurity to Icon
(With a Little Luck in Between)

"For the first 20 or so years, Maker's Mark® bourbon was very well liked by a very select – OK, tiny – group of folks (mostly Kentuckians) who agreed with Dad's ideas about what made a good whisky. But we would have remained a local phenomenon had it not been for a fellow from the Wall Street Journal who paid us a visit in 1980, then decided to write about it on the front page. Never underestimate the power of dumb luck."

— Bill Samuels, Jr.
Former President and Chairman Emeritus

THE WALL STREET JOURNAL.

© 1980 Dow Jones & Company, Inc. All Rights Reserved.

VOL. LX NO. 204 ★★   MIDWEST EDITION   FRIDAY, AUGUST 1, 1980   (H)   35 CENTS

Panicky Pols
More Democrats Fear An Anti-Carter Sweep, But Who Else Is There?

President Slips, Yet Kennedy And Jackson Have Flaws; Muskie, Mondale Demur

—

Convention Could Be Bloody

By James M. Perry And Albert R. Hunt
Staff Reporters of The Wall Street Journal

WASHINGTON—"Free the delegates!" stormed Edward Bennett Williams, the well-known lawyer, at a press conference yesterday in a House office building calling for an "open" Democratic convention.

Over on the Senate side of Capitol Hill, Democratic Sen. Edward Kennedy and independent presidential candidate John Anderson suddenly called a joint press conference to hint that Rep. Anderson might drop out of the race if the Democratic convention nominates someone other than President Carter. Meanwhile, Democratic Senators rushed from one meeting to another, wringing their hands over what might happen to them in the election this fall.

And hovering over all of them is the rollicking figure of Billy Carter, the President's bad-boy brother, who yesterday said a top Justice Department official—who had accused him of lying—is "full of s—."

Thus the Democrats' political panic spreads, just 10 days from the opening of their national convention in New York City. Hour by hour, fed by fire-breathing rhetoric on all sides, the hullabaloo is growing. It promises to make the convention stormier than most, maybe even rivaling the 1968 firestorm in Chicago.

The Real Issue: Jimmy Carter

Behind the rhetoric, the real issue is President Carter himself. His ratings are the lowest the pollsters can remember, plunging to the bottom of their charts over his brother Billy's Libya connections. Some Democrats fear he could be on his way to a crushing defeat in November. Maybe more important is the growing realization among many congressional Democrats that they could be swept out of office along with Mr. Carter.

So the remedy, in the view of these Democrats, is to open the convention and allow the delegates to make another choice. Sen. Kennedy thinks the convention, free to vote its conscience, might turn to him, wishful as that thinking may be. Mr. Williams seems to be thinking of still another possibility, perhaps secretary of State Edmund Muskie, one of his oldest friends.

It still isn't likely that Mr. Carter can be denied the nomination he thought he had locked up in the caucuses and the primaries. The problem for the Democrats who are terrified over the prospects of the Carter ticket is finding someone else.

Many congressional members think Sen. Kennedy, who will take 1,250 delegates to the convention, could be as bad for them as Mr. Carter.

What About Jackson?

Mr. Kennedy is said to have told Sen. Henry Jackson of Washington that if he fills, the next choice should be Mr. Jackson. But many of Mr. Kennedy's supporters are old-line liberal Democrats, and most of them would have trouble swallowing Mr. Jackson's hard-line defense and foreign-policy positions.

That seems to leave Secretary of State Muskie. Mr. Muskie says he supports Mr. Carter "all the way." But he doesn't quite say he wouldn't accept the nomination that he sought so hard in 1972. It would be difficult for Mr. Carter's own Cabinet official to be part of such a disloyal proceeding. It would be even more difficult, of course, for Vice President Walter Mondale, another possibility. Rep. Morris Udall of Arizona, still another, says he would head for the Mexican border if he was nominated.

So the odds remain with Mr. Carter, who will take 2,000 delegates to the convention, 300 more than he needs to win the nomination. "He's been hurt badly," says one party professional, "but he's still the heavy favorite."

Desperate to Dump Carter

Many of the people who are talking most loudly about denying President Carter the nomination — mostly junior Democratic House members—aren't even delegates to the convention. But they are keenly interested in keeping their jobs, and they desperately believe that the only way for them to stay in office this fall is for the delegates to dump Jimmy Carter.

Here is why, according to Joe Rothstein, a Washington-based political consultant who is managing the campaigns of 11 House Democrats and three Senate Democrats.

"Carter at the top of the ticket costs all my people five to 10 points," he says. "I think it's going to be 1932 in reverse" (when the Republicans, led by Herbert Hoover, were routed).

But is it practical to "dump" the President of the United States?

"Is it practical," Mr. Rothstein replies, "to nominate a President with a 77% negative rating, a President who's two to one behind his Republican opponent, a President who has a growing scandal involving his brother? Is it practical not to dump him?"

That's the sort of talk that is spreading across Washington. Polls are coming in from around the country, and some of them show Mr. Carter running third, behind both Ronald Reagan and Mr. Anderson. In one rural House district, Mr. Rothstein says, a poll shows Mr. Carter collecting less than 15% of the vote. "Carter isn't coming back," he says. "I can't think of anything I could do for my people that would be more important than getting Carter off the ticket."

If Mr. Carter really is going to be
Please Turn to Page 9, Column 1

What's News—
* * *
Business and Finance

CHRYSLER reported a record loss of $536.1 million for the second quarter and a 33% sales decline to $2.12 billion. The deficit is close to what the auto maker had estimated. Separately, the Chrysler Loan Guarantee Board gave final approval to $300 million more in loan guarantees for the company.
(Story on Page 3)
* * *
Auto makers, attempting to clear dealers' lots for 1981 models are said to have trimmed 40,000 units from earlier production schedules for the third quarter. That reduction would represent a 20% drop from depressed 1979 levels.
(Story on Page 3)
* * *
Gasoline prices are beginning to fall slightly because of a growing glut of fuel. Among the major companies lowering prices are Cities Service, Texaco, Gulf Oil and Conoco.
(Story on Page 2)
* * *
Farm prices surged 5.2% last month, the largest increase since 1974, as drought fears helped propel average prices for soybeans, corn, hogs, cattle and broiler chickens.
(Story on Page 2)
* * *
New factory orders declined 0.5% in June to $138.21 billion, after falling 3.1% in May. The June drop was the smallest in five months, producing further evidence that the pace of the recession is easing.
(Story on Page 2)
* * *
Reduced FCC regulation of certain small communications carriers is expected to be cleared today by the commission. The proposals would give the carriers greater rate-setting flexibility.

AT&T's entry into the data-processing field and the deregulation of many aspects of telecommunications would be permitted under a bill approved by the House Commerce Committee. The measure faces hurdles in the House and Senate.
(Stories on Page 7)
* * *
Standard Oil of California said its Chevron unit plans to proceed with a $1 billion modification of its Pascagoula, Miss., refinery that will enable the facility to process heavy, higher-sulphur crude oil.
(Story on Page 7)
* * *
Petro-Canada, the Canadian state oil company, signed a formal contract to buy the bankrupt oil refinery at Come By Chance, Newfoundland, for as much as $237 million (Canadian). Financier John Shaheen is expected to challenge the contract in court.
(Story on Page 4)
* * *
Alberta announced domestic petroleum price increases that were within a range acceptable to the federal government, thereby averting a threatened confrontation over oil pricing. The action won't affect export prices. Canadian crude will rise $2 (Canadian) a barrel to $16.75.
(Story on Page 4)
* * *
Canadian Paperworkers Union members approved a two-year contract with Abitibi-Price, the world's largest producer of newsprint. The pact is expected to serve as a pattern at other companies.
(Story on Page 4)
* * *
U.S. Steel's planned sale of its cement division to Lehigh Portland Cement for $138 million was postponed because of FTC opposition. The commission said the transaction would violate antitrust law.
(Story on Page 6)
* * *
Interest rates soared again in money and capital markets and prices plunged amid fears that the Fed has tightened its credit reins. Many economists concluded that the Fed is willing to accept a rise in the key rate on so-called federal funds.

Mortgage rates were raised by some savings and loan associations in California to 12¼% from 12⅛%. The increases reflect higher yields on mortgages accepted by the Federal Home Loan Mortgage Corp.
(Stories on Pages 7 and 14)
* * *
Markets:
Stocks: Volume 54,610,000 shares. Dow Jones industrials 935.32, off 0.86; transportation 312.01, off 2.46; utilities 111.38, off 0.62.
Bonds: Dow Jones 20 bonds 72.12, off 0.57.
Commodities: Dow Jones futures index 447.23, up 4.39; spot index 436.95, up 0.94.

TODAY'S INDEX

Annual Meeting Briefs		Dev'l. Agency Quotes	
Commodities		Int'l. News	
Corrections		Money Rates	
Credit Markets		Securities Markets	
Dividend News		Tax Exempts	
Earnings Digest		Treasury Issues	
Editorials		Who's News	
Financing Business			

World-Wide

CARTER WILL TESTIFY after the Democratic convention, Senators indicated.

As a special Senate panel began its inquiry into ties between Libya, Billy Carter and his brother's administration, Chairman Birch Bayh (D., Ind.) said more information is needed before calling the President. The White House said the President plans to send Congress a report Monday, before Democrats gather in New York Aug. 11.

In Americus, Ga., Billy Carter vehemently denied receiving official cables on Libya. He also disputed Wednesday's charges by a Justice Department official that he lied to hide Libyan payments. A presidential aide again said the administration "will be prepared and eager to respond to any questions."

Senate panelist Strom Thurmond (R., S.C.) said he wants to know why the President "permitted his brother even to deal with a foreign country."

The White House released copies of seven State Department cables praising Billy Carter's 1978 trip to Libya. A spokesman said the reports were the ones discussed by the President and Billy Carter. Most of the material was made public a year ago, the official added.
* * *
ANDERSON HINTED he may drop his presidential bid if Carter is dumped.

The independent candidate said he opposes a choice between the President and Ronald Reagan, rather than the two-party system. Anderson spoke after a meeting with Sen. Edward Kennedy, who trails Carter in pledged Democratic delegates. But Kennedy aides said they are gaining support daily for releasing delegates' pledges.

In Washington, attorney Edward Bennett Williams took the helm of an "open convention" group, but denied he wants to get rid of Carter. Stressing moral issues, he said delegates otherwise could be "held hostage to a tyrannical rule." Sen. Warren Magnuson (D., Wash.) also urged the President to free delegates.

Junior House Democrats have been at the front of open-convention efforts. They feel that anti-Carter sentiment could hurt an entire Democratic ticket headed by the President.
* * *
A lame duck session of Congress is a "near certainty," House Speaker Thomas O'Neill (D., Mass.) said. Passage of a fiscal 1981 budget and a possible tax cut are expected to be the major projects. Holding the first postelection session since 1974 also would allow an early October recess for campaigning.
* * *
Reagan reported 1979 income of $515,878, slightly more than half of which went for taxes. While the Republican nominee claimed deductions as small as $15 for depreciation of a fan, the returns didn't appear to include any targets for Democrats. Reagan didn't release tax data during his GOP campaign, claiming privacy.
* * *
Israel's justice minister resigned, but said he will continue to support Prime Minister Menachem Begin. Shmuel Tamir explained that his minority coalition party was over-represented in the cabinet. In Egypt, officials discussed responses to Israel's Wednesday passage of a law that all of Jerusalem is its capital.
* * *
A women's action program was approved 94-4 at the end of a UN-sponsored Copenhagen meeting. The U.S. opposed the plan because of "abhorrent attacks against Israel," delegate Sarah Weddington said. Sexism was condemned after Third World and Soviet-bloc nations won a footnote saying the term applies only in the West.
* * *
Efforts to ban nuclear tests of weapons "have come far," the U.S., Britain and the Soviet Union announced in Geneva. An accord in principle permits on-site inspections, which can be denied if the host country provides reasons. Conference sources said the three years of talks still leave major work to be completed.
* * *
Closing of judicial proceedings could be subject to tighter limits, the Justice Department proposed. The guidelines require agency permission and a showing that closing is "essential to the interest of justice." The Supreme Court has favored open trials but has allowed judges to close pretrial hearings to the public.
* * *
Iran executed 11 civilians for alleged involvement in a military plot to kill Ayatollah Khomeini, Tehran radio reported. Close to 30 army officers have received the death penalty as well. In a Parisian newspaper interview, Foreign Minister Sadegh Ghotbzadeh said delays in forming a government block his plans to resign.
* * *
Havana welcomed Mexico's president, Jose Lopez Portillo, who arrived for three days of talks. The Mexican delegation is expected to agree to supply Cuba with Mexican technology for oil exploration.
* * *
Turkey won a $4 billion pledge from Swiss, British and West German companies for financing of a dam and hydroelectric complex on the Euphrates River. Turkish officials said the Ataturk project will produce 10 billion kilowatt hours of electricity annually and irrigate two million acres.
* * *
A sale of jet engines to be used in Italian construction of Iraqi warships was approved by the State Department. Israel and some Congressmen have objected to the $11.2 million accord for General Electric equipment.
* * *
The U.S. flag won't be used at Sunday's closing ceremonies of the Moscow Olympics, the International Olympic Committee announced. The White House had protested plans to include the flag, noting the American boycott to protest the Soviets' Afghan invasion.

Hourly Earnings

AVERAGE HOURLY PAY of factory workers in June rose to $7.18 from a revised $7.13 the preceding month, the Labor Department reports.

Maker's Mark Goes Against the Grain To Make Its Mark

* * *
Bourbon Distiller Is a Model Of Inefficiency by Choice; No Case for Fidel Castro

By David P. Garino
Staff Reporter of The Wall Street Journal

LORETTO, Ky.—Maker's Mark Distillery has made its mark by going against the grain.

In producing its premium-priced Maker's Mark bourbon, it continues to use an intricate six-year aging process and a small bottling line that are models of inefficiency. It distills only 19 barrels of bourbon daily, compared with hundreds distilled by other producers. Its ad budget is a meager $1.2 million a year.

But most remarkably, its volume of business has more than tripled, to about 150,000 cases a year, in the past 10 years, while the overall bourbon industry's sales have slipped 26%, to 23.7 million cases.

With its growing reputation for high profitability despite its antiquated production system, Maker's Mark is understandably viewed with envy and lust by some other distillers and conglomerates. But when Maker's Mark is approached by a suitor, says T. William Samuels chairman, "I just won't talk to them." He represents the fourth generation of his family in the distillery business, and he wants the company to remain in the family.

Suitors Not Suitable

So potential suitors are left drooling on the sidelines. "If we had it in our stable," says one major distiller, "we'd promote the hell out of it" to fulfill its potential. But, the executive adds with a sigh, "Maybe they're happy doing what they're doing."

They are indeed. "We'll be satisfied with a steady 8% to 10% yearly growth, says the 70-year-old Mr. Samuels.

Competitors, meantime, will have to keep admiring Maker's Mark from afar. James "Buddy" Thompson, chairman of Glenmore Distilleries Co., observes, "Bill Samuels started from scratch and established a brand of superior quality with a fine image. It's a textbook case of superior marketing."

John Heilmann, who heads Norton Simon Inc.'s liquor operations, observes, "I'll pay Maker's Mark the consummate compliment: If I'm in a bar and it doesn't have any of our brands, Maker's Mark is my first choice." And he would probably pay a premium because Maker's Mark retails in Kentucky for $8.75 for three-quarters of a liter, against $5.70 for the popular Jim Beam bourbon.

Maker's Mark aims to keep its small cadre of fans loyal. And that, says Mr. Samuels's 40-year-old son, William Jr., president of the distiller, means "zero compromise on the quality of our whisky."

Unimposing Distillery

Nestled amid the rolling hills at "Happy Hollow" in this town of 1,000 about 60 miles from Louisville, the tiny Maker's Mark distillery is unimposing. It doesn't disclose its volume, but the estimate of 150,000 cases a year by industry sources compares with about 2.7 million cases turned out by the premium-priced Jack Daniel's, distilled by Brown-Forman. Total employment of Maker's Mark is 56.

Maker's Mark simmers its grain mash in cypress tanks for four hours, against the standard industry practice of half an hour. The whisky, stored in oak barrels, is rotated through its six-year aging process, starting at the top of a six-floor warehouse, and moving from warmer temperatures there to the cooler bottom floors.

"This is terribly labor intensive," says Sam K. Cecil, vice president for production and a 43-year industry veteran. "But that's the best way to make bourbon."

The production line, similarly, handles only 40 bottles a minute, compared with hundreds for competitors, mainly because the caps are sealed with a distinguishing hand-dipped red wax. An inspector with a magnifying glass examines bottles for imperfections and impurities.

Maker's Mark doesn't have a field sales force. Marketing is done mainly by Mr. Samuels Jr., Gus Silliman, senior vice president, and James Conn, national sales manager. "The personal touch" has helped make the brand a success, says Ralph (Bud) Baldwin, Maker's Mark Chattanooga distributor. Not long ago he and Mr. Samuels Jr. in one day visited 20 of his 54 accounts. "It isn't every day that the president of a company walks through a retailer's front door," Mr. Baldwin observes.

When the bourbon was first marketed in 1958, the elder Mr. Samuels tried to establish a classy image, and his first ad was a two-page spread in New Yorker magazine. Today, its advertising budget is targeted at newspapers, regional editions of magazines such as Time, Playboy and Penthouse, and specialty publications including Southern
Please Turn to Page 9, Column 5

World Bank Affiliate Approves $67.2 Million In Loans to 4 Nations

By a Wall Street Journal Staff Reporter

WASHINGTON — The International Development Association, a World Bank affiliate, approved $67.2 million in loans to four countries.

Nepal will get a $27 million credit for public water-supply projects in urban areas, while Tanzania will receive a $25 million loan for the construction of schools and other education facilities.

The IDA also approved two smaller loans—$7.7 million to Burundi and $7.5 million to Rwanda—to expand the telephone and telex communications systems in both countries.

All of the IDA credits will be interest-free, except for a small annual administrative charge, the Department reports.

Allis-Chalmers Forms A Venture in Argentina

By a Wall Street Journal Staff Reporter

MILWAUKEE — Allis-Chalmers Corp. said it formed a venture with Argentina's national shipyard, Astilleros y fabricas Navales S.A., to produce hydraulic turbines and other heavy equipment.

The industrial equipment maker said it hasn't decided how much it will invest in the venture. The venture, called Afne-Allis, plans to build a manufacturing facility about 30 miles south of Buenos Aires.

Allis-Chalmers said last month that it was the apparent low bidder to manufacture 20 hydraulic turbines valued at a total of $172 million for a hydroelectric project on the Parana River between Argentina and Paraguay. Allis-Chalmers said that if it is awarded that contract, some of the parts would be made at the planned plant.

Living and Louisville Lawyer.

Its ad tag line is: "It tastes expensive . . . and is." Typical of its copy: "For those who ask how good a whisky is. Rather than how much."

Origins of the Species

The elder Mr. Samuels got into the distilling business naturally enough. A forebear brewed up whisky for Washington's militia during the Revolutionary War, and the family has seldom been far from that business ever since. Kentucky's leading distillers, the Bearns, the Samuelses and John Shaunty, who owned Early Times, lived on the same street, known as "Whisky Row" in Bardstown, about 15 miles from here. Mr. Samuels Jr. recalls sitting on the knee of the legendary Jim Beam, listening to tales about the industry.

The father of Mr. Samuels Sr. built the T.W. Samuels brand into a leading whisky early this century, but after Prohibition, the majority control was sold. Mr. Samuels Sr. left the industry in 1943 and sold bottling machinery for 10 years.

After the Korean War, many distillers were in trouble because they had built up big inventories in anticipation of government restrictions that never materialized. Mr. Samuels Sr. bought out one bankrupt distillery, picking up 202 acres for only $80,000.

Finding the family's traditional plaster unwilling to back his venture, Mr. Samuels went to National Bank of Louisville. Hubbard Buckner, retired senior vice president of the bank, recalls, "Since Prohibition, I had never known any bourbon maker who put out a legitimate product to fail. I knew Bill Samuels would have a superior bourbon."

Starting Anew

Mr. Samuels says with a broad smile, "I'm probably the only distiller who threw the family's formula into the trash can. T.W. Samuels was a good whisky but it wasn't unique." (It has long since disappeared from the marketplace.)

His new formula substituted more-expensive wheat for rye to give a smoother taste. Mr. Samuels draws a comparison between wheat and rye bread, with the latter having more bite. Among others, David Kay, president of Pet Inc.'s liquor retail chain, thinks the taste is different. "Maker's Mark is so mellow, an excellent sipping whisky," he says.

The naming of the brand was "as un-Madison Avenue as could be," Mr. Samuels recalls. While discussing possible names at Mr. Samuel's Bardstown home, an advertising executive noticed the craftsman's sign, or touchmark, on the bottom of a pewter mug. Hence, the name "Maker's Mark." The bottle features a circle encompassing an "S" for Samuels, a "IV" for fourth-generation distiller and a star for Star Hill Farm, the distillery property.

Distilling from scratch had a distinct disadvantage Mr. Cecil notes. "We couldn't sell drop one for six years," and before the first sale was made in 1958, "we were sitting on more than $1 million of inventory."

Recent Profits

It took 10 years to erase the initial deficits, but cumulative profit the past three years has exceeded all of the first 24 years, says Mr. Samuels Sr.

Politicians hereabouts show an affinity for Maker's Mark, says Mr. Gregg, the Louisville distributor. "It wouldn't be graceful to name them," he observers. But Tim Lee Carter, a Congressman from Kentucky, isn't bashful about his preference. Not long ago Dr. Carter on the House floor praised Maker's Mark as "that incomparable bourbon of bourbons." He had just won from Indiana Congressman John Myers a bottle of Maker's Mark when the University of Kentucky beat Indiana in a basketball game.

Cuban Premier Fidel Castro also appears to be a fan. When Frankfort, Ky., Mayor John Sower and his wife, Phyllis, went to Havana on a national league of cities tour in 1978, Mr. Castro greeted the group at a governmental palace. When he learned that the Sowers were from Kentucky, he asked them to send him a bottle of Maker's Mark. Says Mrs. Sower, "We didn't — we think he's an S.O.B."

THIS COLUMN REPRINTED FROM PAGE 9

Onward & Upward
Led by Food Prices, Inflation Seems Likely To Maintain Fast Pace

—

Hopes That Consumers Soon Will Get a Break Decline; Drought, Other Ills Cited

—

This Time, Oil Isn't to Blame

By Lindley H. Clark Jr.
Staff Reporter of The Wall Street Journal

Economists who thought consumers were about to get some price breaks now see mainly some broken hopes.

Until recently, analysts had expected the consumer price index to rise much more slowly in the summer and early fall. They had envisioned perhaps an annual rate of gain as low as 5% or 6% in the current quarter, down from the 15% rate in the first half.

Now they say an easing still is likely, but think it won't be that dramatic. "I look for consumer prices to rise at an annual rate of 9.4% in both the third and fourth quarters," says Lacy H. Hunt, chief economist of Philadelphia's Fidelity Bank.

That's only the near-term bad news. The 1981 bad news: Most forecasters think that the consumer price index will be rising at an annual rate close to 10% next year. Moreover some say, the rate may be accelerating before the year is over.

Surging Food Prices

The price outlook has turned gloomier largely because of the recent surge in food prices. The latest evidence of this surge was the Agriculture Departments' report yesterday that between mid-June and mid-July farm prices rose 5.2%, one of the sharpest one-month increases on record. Without compounding, that's an annual rate close to 62% (see story on page 2).

"Clearly, the drought and hot weather have had a dramatic effect on agricultural prices and commodities inflation," says Donald Ratajczak, director of the economic-forecasting project at Georgia State University.

And analysts reason that continued high rates of inflation, in turn, will tend to keep interest rates close to current levels. Mr. Hunt suggests that the banks' prime rate, the fee charged their best-rated borrowers, "may go back up again." The prime rate, which was reduced to 10¾% by some banks last week, may rebound to 11% or 11¼% in the next few weeks, Mr. Hunt thinks.

High inflation and higher interest rates would tend to hold down the recovery from the current recession—a recovery that most analysts expect to begin around year-end—to a slow and modest pace. If consumer prices rise at an annual rate of about 10% next year, Lawrence Chimerine of Chase Econometrics sees as unlikely any recovery in real incomes—incomes adjusted for inflation. Moreover, he expects much of the inflation next year to be in food prices, and he says costlier food tends to depress consumer spending especially severely by reducing discretionary income.

Undermined Hopes

A rise in interest rates would further undermine the earlier optimism about this summer's price trends because those hopes had been based partly on this spring's sharp drop in home-mortgage interest rates, which figure importantly in the consumer price index. Many economists also had lowered their inflation forecasts when Congress refused to pass the Carter administration's oil-conservation fee, which would have increased gasoline prices.

But now analysts such as Sam Nakagama, chief economist at the brokerage firm of Kidder, Peabody & Co., expect the effects of the heat and the drought in the Midwest and the South to linger for some time. A government survey showed, Mr. Nakagama says, that "hog producers were planning to reduce sow farrowings by 8% in the June-November period. Because of hot weather in the farm belt over the past month, the pig crop actually may decrease by 12% or so because of poor conception rates and stillbirths." A reduced supply would point to higher prices in the first half of 1981.

The heat wave has killed millions of broiler chickens and pushed up their prices. It also has killed more than one million breeding hens, Mr. Nakagama adds, and that will cut broiler production in early 1981. The heat and the drought have led ranchers to sell cattle earlier than usual and thus helped to hold down beef prices this summer. But, Mr. Nakagama notes, "these effects should disappear by the fall."

Some Other Problems

All in all, Mr. Nakagama figures that food prices will add two percentage points to the consumer price index over the next year. And contributing to the dismaying outlook on inflation are difficulties with monetary policy, productivity trends and labor costs.

That outlook has deteriorated even though economists acknowledge that the consumer price index for July, due out on Aug. 22, may rise "only" 8% or even a bit less. They expect this relative moderation in some of the lower mortgage rates, softening prices of gasoline and the fact that many food-price increases haven't yet reached the consumer level.

Discussing monetary policy in congressional testimony last week, Federal Reserve Chairman Paul A. Volcker reaffirmed the Fed's determination to continue to slow the growth in the nation's money supply. But neither he nor many private economists expect the Fed's program to have much effect on inflation for some time.

Mr. Volcker commented that "we are in the process of seeing" inflation come down to the "core" rate of about 9% or 10%. "That core rate is roughly determined by trends in wages and productivity," he told
Please Turn to Page 9, Column 6

The article read 'round the world: Maker's Mark is featured on the front page of The Wall Street Journal, the first time a small, family-owned company had been so recognized.

We were very pleased on August 1st to find a story about our little family distillery on the front page of The Wall Street Journal.

As a result of the story, we're getting calls from people all over the country who are "suddenly" interested in buying Maker's Mark.

And as much as we'd like to accommodate all the inquiring public, we're concerned that we can't. Quality is what makes Maker's Mark special. And if we made much more than we did, well, it just wouldn't be the Maker's Mark you read about.

If our special bourbon whisky isn't available where you live, you might need a little perseverance. If your local retailer doesn't have it, he can order it for you.

Or, if you'd prefer, write us at Maker's Mark. We'll get you started on the right avenue toward finding this one-of-a-kind whisky.

It tastes expensive™
...and is.

Bill Samuels Jr., President

Bill Samuels Sr., Chairman

Maker's Mark Distillery, Loretto, Ky. 40037 • Ninety Proof, Fully Matured

The Home Place of Bourbon Making

"My mom knew what she was doing when she insisted that Dad set aside some of the profits from our whisky to turning the distillery into the kind of place people would want to visit and tell their friends about. Being able to walk around the place, smell the mash cooking, and talk with the people who actually made the whisky – that's always been a big part of what makes Maker's Mark special to our fans. And as long as I have anything to say about it, it always will."

— Bill Samuels, Jr.
Former President and Chairman Emeritus

The historic Burk's Spring distillery, reborn as the home of Maker's Mark thanks to my mother. The functioning arboretum was created with help from the University of Kentucky and features live species native to the region.

Visitors to the distillery know they are getting close to their destination when they begin seeing the classic dark brown rickhouses dotting the landscape in and around Loretto. Here, a rickhouse is silhouetted against another beautiful Kentucky sunrise.

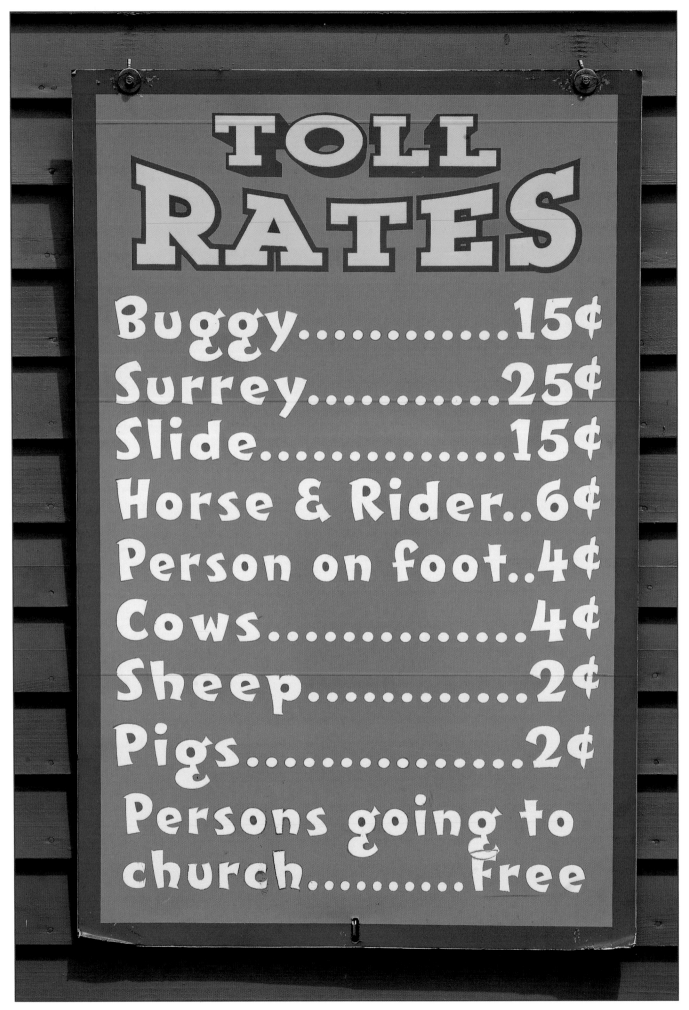

A sign posting the rates to use the private road short-cut through the distillery property. Owners often collected tolls to help offset the cost of maintaining the roads.

The Toll House, once used to collect tolls, now serves as a café for visitors to enjoy a snack while visiting the distillery.

A sign with the iconic Maker's Mark® bourbon logo welcomes visitors to the distillery.

1950's farm building that was often used as a bill board.

Whisky Creek flowing into the distillery property once powered the gristmill.

The Distiller's House was the original family dwelling. Later it became the home of the Master Distiller who was needed on-site 24 hours a day in case of production problems. It now serves as the Welcome Center and starting point for the distillery tours.

The distillery boasts the world's only trapezoidal bridge, which is one more thing that makes this place so unique. Here it is seen in various seasons of the year.

Opposite: *The risk of fire is high and always present. Back in the day, distilleries maintained their own volunteer fire departments to respond quickly to a fire in these remote areas.*

Boys and their toys… The Firehouse serves as home of the restored 1928 fire truck.

These lucky visitors get a special introduction to the Maker's Mark heritage by former President and Chairman Emeritus Bill Samuels, Jr., at the beginning of their tour of the distillery.

An authentic record of Robert Samuels paying tax on whisky processed during year one in Bourbon County hangs in the Distiller's House.

A scroll welcoming visiting Ambassadors to the distillery holds a place of prominence in the Distiller's House.

Opposite: *A frosty sign complete with the Maker's logo atop provides directions and a warm welcome to visitors.*

Pewter models and prototype bottles also hold a place of honor in the retro 1950's kitchen in the Distiller's House.

Left: *Various grains used in the search for the "perfect" whisky. Hundreds of loafs of bread were baked in the family kitchen with different proportions and ingredients as part of this research and development process.*

Below: *A look at early experiments of the signature Maker's Mark® bourbon red wax seal is on display in the kitchen of the Distiller's House. Margie Samuels demonstrated her vision for hand dipping by heating wax in her deep fryer cooker.*

Maker's Mark Distillery is designated as a National Historic Landmark.

Rickhouse "A", the oldest storage building at Maker's Mark, dates back to 1889.

Previous spread: *The Quart House, circa 1889, has been designated the oldest standing retail package whisky store in America.*

This page and opposite: *Artifacts abound inside the Quart House.*

Whisky was sold in the Quart House directly from the barrel, uncut and unfiltered.

Numerous old time bottles and stencils in the Quart House.

An overall view of the interior of the 1800's retail outlet.

Right: *No computers here, records for tax purposes were kept by hand in ledgers.*

Far right: *Prior to the introduction of glass bottles, whisky was dispensed from the barrel into jugs which were brought by the customers.*

Previous spread: *The Barrel Wagon reminds us how whisky was transported in the 1800's and provides a great backdrop for photos during a visit.*

Below and opposite: *Views of the Maker's Mark Barrel Wagon in various seasons of the year.*

The iconic Maker's logo adorns a wrought iron fence on the property.

The old mill stone, which was the heart of the water-powered gristmill, rests outside the Stillhouse.

Steam rises into the sky over the Stillhouse as another batch is started.

Various exterior views of the Stillhouse.

Detail views of the exterior of the Stillhouse. The color scheme, dark brown for the bourbon, cream for the label and red for the wax, was developed by Margie Samuels.

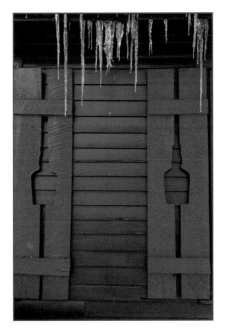

Another of Margie's ideas, the shape of the Maker's Mark® bourbon bottle is cut into all of the shutters. Bill, Sr. pointed out that this was not original to the buildings, but you see who won the debate.

Previous page: *Christmas is a special time at the distillery. Every window gets a wreath and bow.*

Following spread: *The spring fed "Maker's Lake" holds all the water used in the production of Maker's Mark®.*
The limestone shelf in this area of Kentucky removes the iron from the water and makes the perfect water for bourbon.

Opposite page: The grains used in the production of Maker's Mark® bourbon undergo thorough testing prior to acceptance. No test is more important than the "Smell Test".

Top and center: Tractor trailer loads of grain regularly deliver to the distillery to keep the process running.

Bottom: Operators check the grain to assure it is roller milled to specification. While most distilleries use a hammer mill, Maker's uses a roller mill to maximize the flavor of the grains.

Following spread: Sacks containing stems and other non-grain material that has been cleaned from the grains prior to processing sit in the shadows.

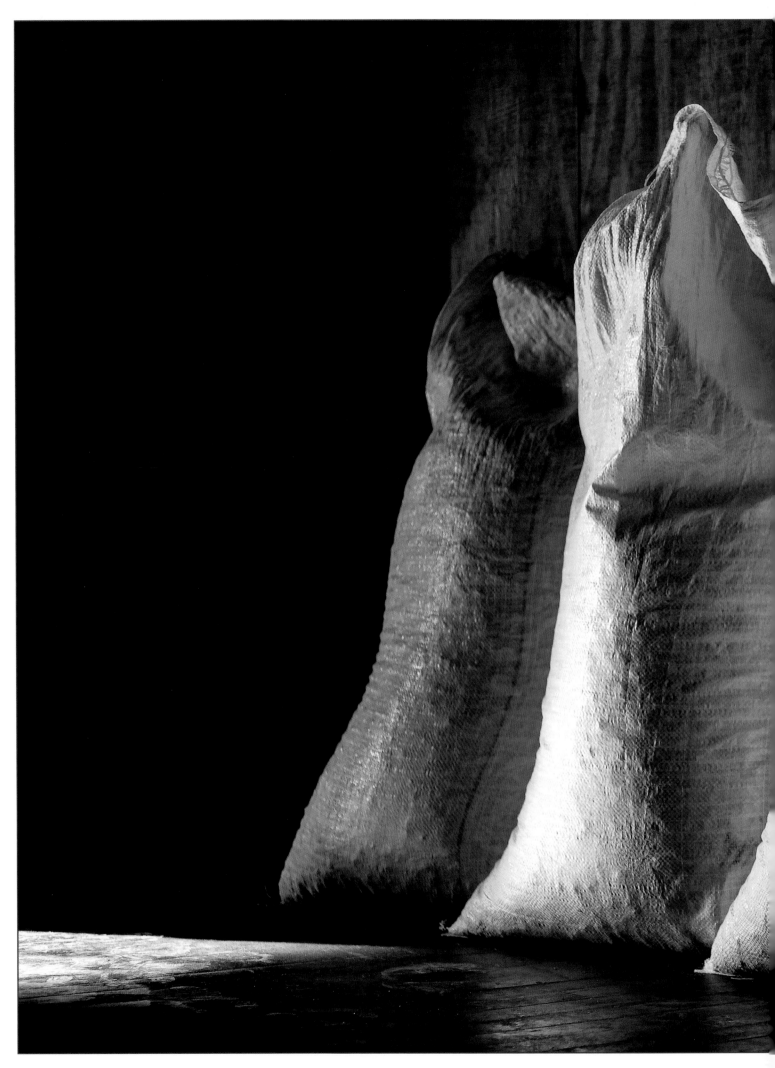

Above: A view of the grounds from a second floor window of the Stillhouse.

Right and opposite: The quantities of the grains in the Mash Bill are weighed by hand using beam scales. No computers here!

The storage bin containing the red winter wheat that makes up part of the Mash Bill.

The "Samuels' Yeast" used in the Mash Bill dates back eight generations and plays an important role in the unique flavor profile.

Opposite and above: *The "Tail Tanks" used in the production of Maker's Mark® bourbon are one of the most popular (and photographed) parts of the tour.*

Keeping these copper kettles clean and shiny is an every day task, but workers still take time to explain the process to visitors on the tour.

83

Samples are regularly drawn from the "Tail Tanks" for evaluation by the Taste Panel.

A bird's eye view of the "Tail Tanks" in all their glory.

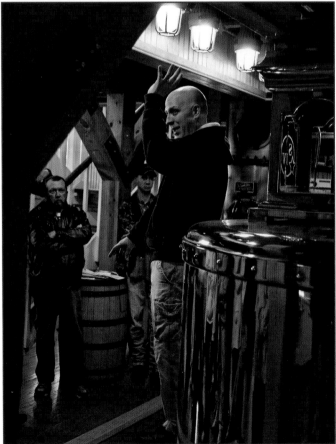

The work must go on. Visitors taking the tour have the opportunity to see the operators in action making Maker's Mark® bourbon.

Next spread: *Another feature unique to Maker's process is the use of "open cookers" rather than the conventional pressure cookers. This results in subtle grain flavors permeating the whisky and adds to its distinctive taste profile.*

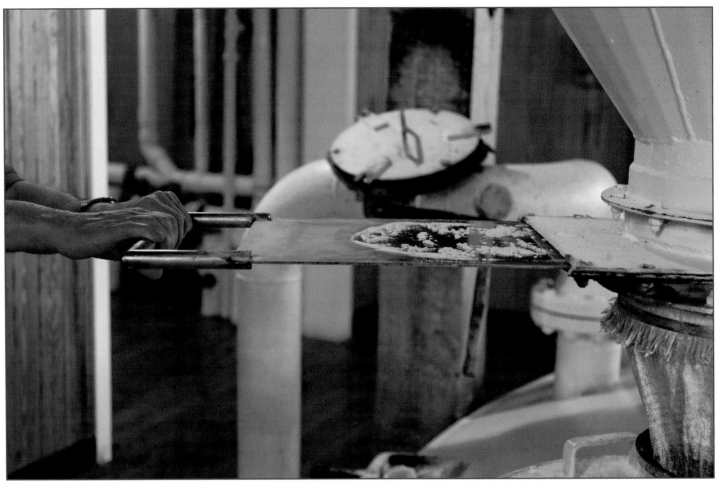
Grains weighed in the scale on the second floor are dropped into the cooker.

Operators strike the side of the hopper with a rubber mallet to make sure all the grain is added to the cooker. It also is a way to signal other operators the hopper is empty and ready to weigh the next ingredient.

New mash is pumped into a tank to begin the fermentation process.

Vats of Maker's™ in various stages of the fermentation process.

The Samuels' Yeast at work. Converting the sugars in the grains to alcohol is a three day process. Many of the cypress boards in these fermentation vats are over 100 years old.

Another batch on its way to becoming Maker's Mark® bourbon ferments in one of eight cypress tanks.

Left: *A view of the grounds from a window in the Fermentation Room.*

Below: *The Maker's Mark® bourbon logo stands as reminder of the history and heritage of the Samuels' bourbon distilling.*

Early morning views of the grounds at Maker's Mark.

Opposite: *Maker's™ White ready for evaluation by the Taste Panel.*

Two of the most sophisticated tools, the human nose and pallet, are used to evaluate Maker's Mark® for quality and consistency.

The Quality Control Lab assures each product is consistent with the Maker's™ standards.

Maker's Mark® is "aged to taste" rather than aged to a specific date. Employees from all departments serve on the Taste Panel on a rotational basis to assure both the quality and taste are maintained. It's not Maker's™ until these folks say its Maker's™. What a job!

The barrel plays an important part in the making of bourbon. Here are some of the vintage tools of the cooperage trade used to build a barrel.

106

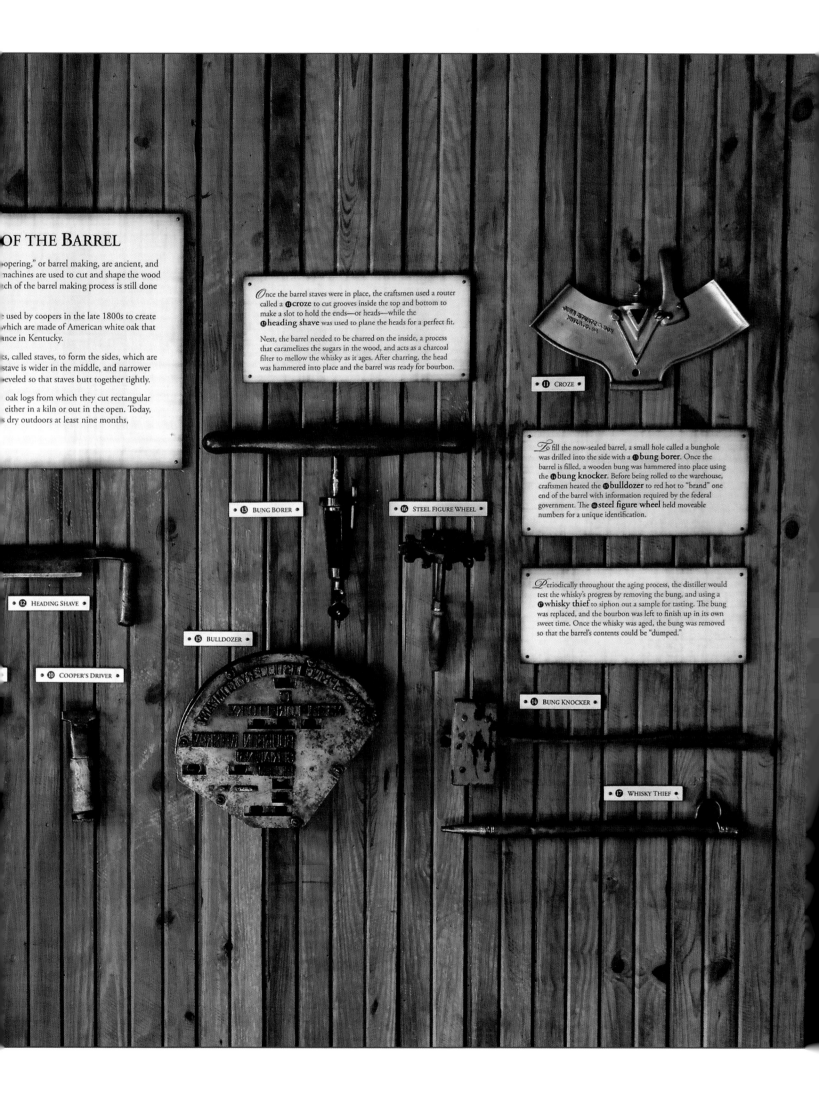

Left: A "Barrel Raiser" adds the final staves to a barrel in the first step of the process. The widths of the staves are alternated to avoid a flat spot in the barrel. Staves are seasoned outdoors for a period of time according to Maker's specifications to remove the tannins from the wood which would adversely affect the flavor of the bourbon.

Below: A raised barrel with a temporary ring affixed and on its way to the steam tunnel.

Opposite: Finished and charred barrel heads ready for assembly. Heads are interchangeable and can be used as tops or bottoms in the final barrel.

A close up view of barrel heads being toasted.

Barrels make a 15 minute trip through the steam tunnel to expose the wood to heat and moisture in order to make the wood pliable prior to bending.

Barrels then pass through a dry fires tunnel in order to maintain the temperature of the wood and aid in bending.

A barrel prior to charring.

Barrels are then charred according to Maker's Mark specifications.

Charring the inside of the barrels adds to Maker's™ unique taste and is also a requirement to be considered "bourbon".

Pressure testing a finished barrel for leaks.

Finished barrels staged and ready for shipment to the distillery.

A view of the inside of a barrel after charring.

Following spread: *An aerial view of the Maker's Mark distillery from the West.*

Mindy, one of the Maker's tour guides, shows a group the inside of a Maker's Mark® bourbon and Maker's 46® bourbon barrel.

Bobby Joe, Rickhouse and Maturation Manager, keeps this part of the operation running smoothly with his years of experience and ever-present smile.

Opposite: *Every barrel is inspected before it is accepted and filled with Maker's White.*

Below: *A Maker's Mark® Ambassador barrel is readied for filling.*

Above: *Barrels are unloaded by hand at the filling station.*

Left: *Barrels arrive from Independent Stave Company for filling on a just-in-time basis.*

After final adjustment for proof, the barrels are filled with Maker's™ White and loaded for shipment to a Rickhouse for aging.

Charles keeps a watchful eye on the filling process…literally.

Pounding the bung to seal the barrel for a trip to the Rickhouse, "One Hit Charlie" as I came to call him, is able to do it with one strike of his mallet....WAY harder than he makes it look... believe me!

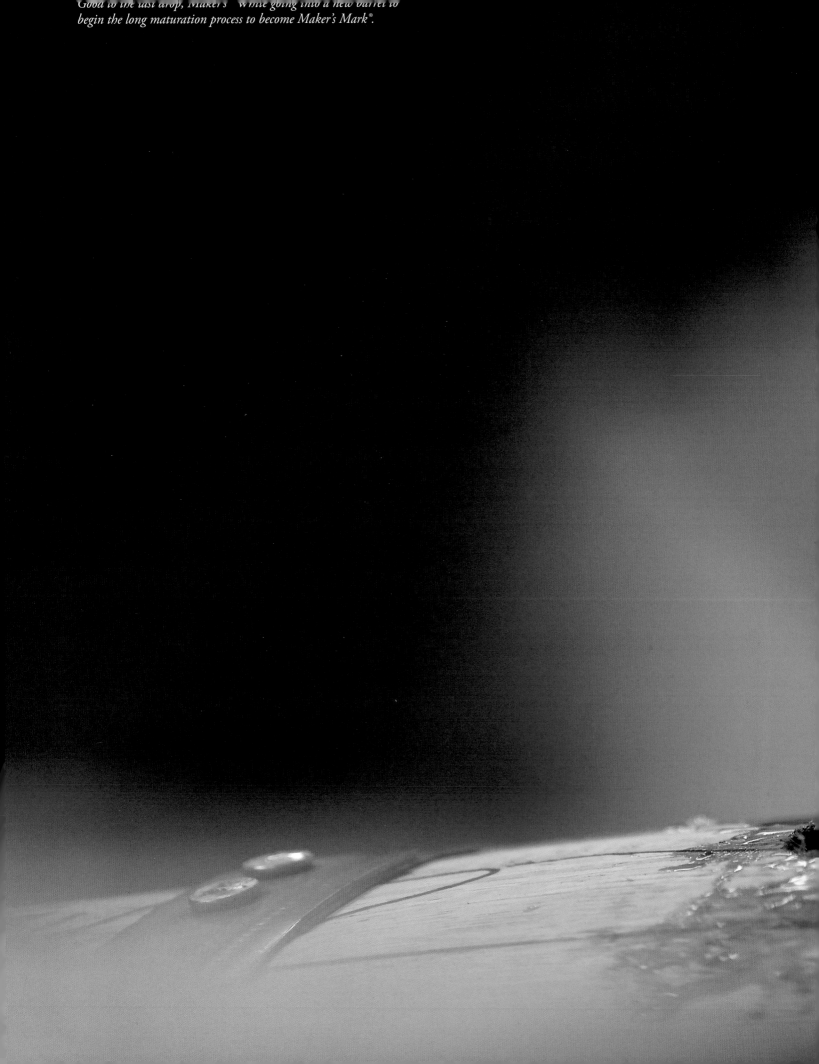
Good to the last drop, Maker's® white going into a new barrel to begin the long maturation process to become Maker's Mark®.

Above: *A look at one of the newer Rickhouses at Maker's Mark. New Rickhouses are being added on a regular basis to keep up with demand.*

Opposite: *Row after row of barrels age in a Rickhouse to one day become Maker's Mark® bourbon.*

Oops! A slight leak between the staves has caused caramelized sugar to form on the outside of this barrel. It looks like the angels got an extra share from this one.

Views of Rickhouse "A" in various seasons of the year.

An example of hand rotating the barrels in the Rickhouses.

Warehousemen are silhouetted in a window of a Rickhouse and flanked by row after row of soon-to-be Maker's Mark®.

Top: *Barrels age peacefully and bask in the sunlight in one of the Rickhouses.*

Middle: *If you look closely, you'll see a plumb line used to indicate the Rickhouse's load is balanced.*

Bottom: *With only natural light available, Warehousemen often use "miner's lights" to help read the lot numbers on the barrels.*

After aging in the upper floors of the Rickhouse for three years, this barrel is ready to be hand rotated to a lower floor.

A Warehouseman loads barrels onto the elevator for their trip to the top of the Rickhouse.

Barrels at the top of the Rickhouse waiting to be rotated down to lower levels.

The sun cascades through the window of Rickhouse "A" on a half barrel bearing the Maker's Mark® bourbon logo.

The passing of the baton. Bill's last barrel and Rob's first barrel rest side-by-side in Rickhouse "A".

A barrel bearing the signature of Chief Operating Officer, Rob Samuels, eighth generation distiller in the Samuels' family.

The aged-to-taste Maker's Mark® is emptied from the barrels and on its way to the bottling process.

A close up view of Maker's Mark® making its long awaited appearance. Vapors, also known as the "Angel's Share", wisp from the barrel as it is emptied.

An exterior view of the Label Room as snow on the trees gives the appearance of cotton balls.

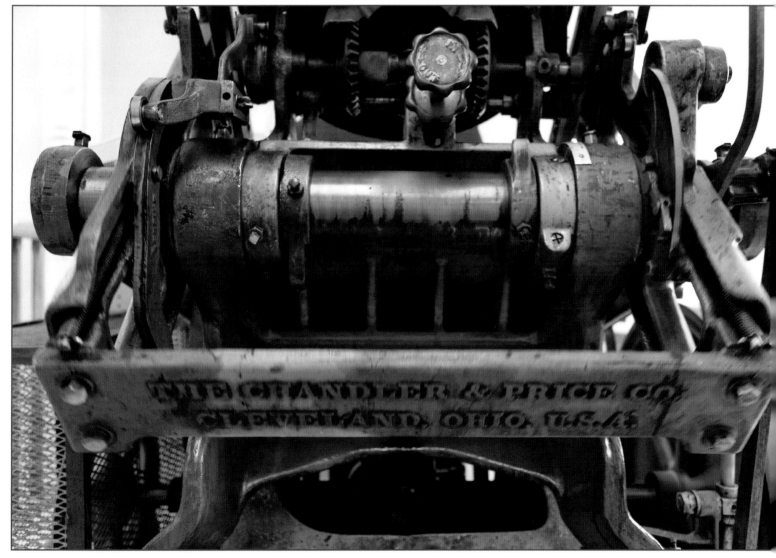

A 1935 Chandler and Price printing press is used to dye cut the Maker's Mark® bourbon labels. After years of faithful service, the original press needed replacing. The search started and an exact replacement was found on the web to insure the tradition continues.

Only four labels are cut at a time, with each and every label being printed and cut on-site at the distillery.

Bottles are rinsed in Maker's Mark® bourbon prior to filling. Water is not used to avoid diluting the final product.

Bottles being filled during the filling process.

Opposite and next spread: *The classic Maker's™ bottle in various stages of being filled.*

Every bottle of Maker's Mark® is hand dipped by actual hands.

The crowning moment in the process comes when each bottle is hand dipped in that signature red wax.

True poetry in motion. Visitors to the distillery can purchase and dip their own bottles in the Gift Shop.

Opposite: *After dipping, the bottles travel through a cooling tunnel to help the wax harden. Even the cooling tunnel sports those unique shutters on its viewing window. From there they continue down line where they are hand packed in cases and readied for shipment.*

From the largest to the smallest, every bottle filled at Maker's gets the same "hands on" treatment.

The first bottle of Maker's Mark® filled on May 8, 1958, and signed by all 18 employees at that time.

The first bottle of Maker's Mark® and the first bottle of Maker's 46® side by side.

Different shapes, different sizes, but all with that signature red wax and commitment to quality.

Bill Samuels, Jr., left, and Rob Samuels, right, hold the two "Firsts" of Maker's Mark®. The pewter piece between them was used by Margie as the starting point for her design of the bottle.

Bill's personal collection of whisky that documents the Samuels family's long tradition in the distilling industry.

Like a proud papa, Bill holds the first bottle of Maker's Mark®.

Even hand surgery can't slow Bill down. Here, he signs a copy of his autobiography for a visitor in the gift shop.

Bill is always at the ready to spread the word about Marker's Mark®.

Below: *Although now "officially" retired, Bill still visits the distillery. Here, some lucky visitors get a first hand explanation of the unique flavor profile and how to best sample Maker's Mark®.*

The Maker's logo embossed on glass window in the gift shop.

At the end of their tour, visitors over 21 years old can sample the various Maker's™ products for themselves in the Tasting Room.

Opposite: *Whisky Creek runs through the distillery's property. The one-of-a-kind trapezoidal bridge can be seen in the background.*

The sound of the flowing water, mixed with the aroma of mash in the air, adds to the tranquility surrounding the grounds at Maker's Mark.

The turn-of-the-century house that was originally the home for the Burks family, now restored and used as the Maker's Mark Club House is pictured in various seasons of the year.

Stone fences, common in this area of Kentucky, abound at the distillery. They also reflect the Samuels family's Scottish heritage.

Following spread: *Complete with rocking chairs, the front porch of the Club House is a great spot to enjoy a sunrise.*

In autumn, the grounds at the distillery come alive in a blaze of color. Margie used the landscaping as a way to showcase the various species of plants and trees native to Kentucky.

The setting sun peaks through one of the rockers on the porch of the Club House.

Farm buildings surrounding the distillery remind visitors this once was a working farm.

Opposite: *The stained glass front door of the Club House is a warm welcome for guests.*

Whisky Creek ends at the back of the property where it empties into Hardin's Creek.

Opposite: *The conference room in the office of Maker's Mark is filled with many artifacts that recall the long heritage and history of the Samuels Family. Here, Bill Samuels, Sr. keeps a watchful eye over his vision of "making bourbon that actually tastes good".*

Another artifact in the conference room is the personal firearm of Frank James (yes…of the "James Gang"). It was presented to Ora Samuels on July 25, 1865 at Samuels Depot, KY by Frank James on the occasion of his surrender officially ending the American Civil War.

A sneak peek at the latest changes at Maker's. Continuing Margie's tradition of fixing the place up, Warehouse A underwent a facelift in the Spring of 2012.

To accommodate the ever-increasing number of visitors to the distillery, Maker's recently expanded the Visitors Center to include three new tasting rooms. If the photos on the wall in the top image look familiar, you're right! Maker's invited the author and other area artists to help depict what sets Maker's Mark® bourbon apart from the crowd.

As the sun sets and we leave Maker's Mark, we vow to return; if not in person, then certainly in spirit when we enjoy their great bourbon.

Index

A
Angel's Share 149

B
barrel heads 108, 110
Barrel Raiser 108
barrels 51, 54, 106, 108, 112, 113, 116, 120, 123, 125, 127, 128, 134, 137, 139, 140, 144, 147, 149
Barrel Wagon 56, 58
Beam, Jim 12
bottling process 147
Bourbon County, Kentucky 40
Bradshaw, Sydina 7
Burks family 183
Burk's Spring distillery 28

C
charred/charring 108, 113, 114, 117
Club House 183, 185, 189, 190
cooker 43, 92
cooling tunnel 163
cooperage 106
copper kettles 83

D
Distiller's House 35, 40, 42, 43
Doe-Anderson 7

F
fermentation 93, 95, 96, 98
Fermentation Room 99
filling process 154
Firehouse 39

G
grain 43, 70, 71, 74, 88, 92, 96
gristmill 35, 60

H
hand dipping/hand dipped 21, 43, 158, 159, 163
Hardin's Creek 192
hopper 92

I
Independent Stave Company 7, 123
iron-free water 8

J
James, Frank 195

L
Label Room 150
labels 153
limestone shelf 68
Loretto, Kentucky 9, 30

M
MacRae-Samuels, Victoria 7
Maker's Lake 68
mash 93, 181
Mash Bill 74, 76, 78

O
old mill stone 60
open cookers 90

Q
Quality Control Lab 103
Quart House 46, 48, 51, 52

R
red wax 21, 43, 159, 169
rickhouse 30, 125, 127, 130, 134, 136, 137, 139, 140
Rickhouse "A" 45, 134, 141, 142
Roberts, Beth 7

S
Samuels, Bill Jr. 7, 9, 11, 15, 16, 19, 20, 23, 27, 40, 142, 170, 171, 172, 173
Samuels, Bill Sr. 8, 9, 11, 15, 20, 21, 23, 27, 40, 67, 195
Samuels Depot, Kentucky 195
Samuels family 8, 9, 170, 185, 195
Samuels, Les 12
Samuels, Margie Mattingly 8, 13, 19, 20, 21, 27, 43, 64, 67, 170, 188
Samuels, Ora 195
Samuels, Rob 7, 8, 142, 144, 170
Samuels, T. William 16
Samuels' Yeast 78, 96
Smell Test 70
staves 108, 132
Stillhouse 16, 60, 61, 62, 64, 74

T
Tail Tanks 81, 85, 86
Taste Panel 85, 103, 105
Tasting Room 177, 197
Toll House 33
trapezoidal bridge 36, 181
T.W. Samuels Distillery 12, 13, 21

V
Vawter, David 7
Visitors Center 197

W
Wall Street Journal 23, 25
Welcome Center 35
Whisky Creek 35, 181, 192
Wickham, Jimmy 7